The Period Positivity Book

THE PERIOD POSITIVITY BOOK

Copyright © Octopus Publishing Group Limited, 2025

All rights reserved.

No part of this book may be reproduced by any means, nor transmitted, nor translated into a machine language, without the written permission of the publishers.

Claire Chamberlain has asserted their right to be identified as the author of this work in accordance with sections 77 and 78 of the Copyright, Designs and Patents Act 1988.

Peer reviewed by Dr Cai Jenny Xiao

Condition of Sale
This book is sold subject to the condition that it shall not, by way of trade or otherwise, be lent, resold, hired out or otherwise circulated in any form of binding or cover other than that in which it is published and without a similar condition including this condition being imposed on the subsequent purchaser.

An Hachette UK Company
www.hachette.co.uk

Vie Books, an imprint of Summersdale Publishers
Part of Octopus Publishing Group Limited
Carmelite House
50 Victoria Embankment
LONDON
EC4Y 0DZ
UK

This FSC® label means that materials used for the product have been responsibly sourced

MIX
Papper | Bidrar till ansvarsfullt skogsbruk
FSC® C018236

www.summersdale.com

The authorized representative in the EEA is Hachette Ireland, 8 Castlecourt Centre, Dublin 15, D15 XTP3, Ireland (email: info@hbgi.ie)

Printed and bound in Poland

ISBN: 978-1-83799-507-3

Substantial discounts on bulk quantities of Summersdale books are available to corporations, professional associations and other organizations. For details contact general enquiries: telephone: +44 (0) 1243 771107 or email: enquiries@summersdale.com.

The Period Positivity Book

Claire Chamberlain

A Guide to Understanding, Owning and Celebrating Your Period

Disclaimer

Neither the author nor the publisher can be held responsible for any injury, loss or claim – be it health, financial or otherwise – arising out of the use, or misuse, of the suggestions made herein. This book is not intended as a substitute for the medical advice of a doctor or physician. If you are experiencing problems with your physical or mental health, it is always best to follow the advice of a medical professional.

Contents

Introduction
6

Chapter One: Period 101
8

Chapter Two: Your Cycle and You
72

Chapter Three: Breaking the Taboo
140

Conclusion
154

Resources
156

Introduction

Welcome to *The Period Positivity Book* – an open, frank, no-nonsense guide to everything relating to your menstrual cycle, your hormones and you.

If you menstruate, then your periods are a completely natural part of your health and well-being. Yet, we're so often made to feel like they're something to be ashamed of – something to be covered up, hidden and not spoken about. This book is here to change that.

Over the coming pages, we'll take a deep dive into all you need to know about your periods. We'll also take an in-depth look at how tracking and becoming familiar with your cycle can not only help you understand your body and emotions more deeply, but can also help you feel more positive, empowered and in control of your life.

Are you ready to embrace your menstrual cycle as something to be celebrated? Then let's get started!

Chapter One: Period 101

Have you ever had questions about your periods (perhaps wondering whether yours are "normal") but felt too embarrassed to ask? Then fear not – this chapter has you covered.

The following pages will explore everything from the timeline of your menstrual cycle and what the colour of your menstrual blood means, to the symptoms of premenstrual syndrome, irregular cycles, conditions that might affect menstruation, period products and more.

Now's the time to empower yourself with knowledge about your body and your cycle, so you can learn how to work with your hormones and feel your best, whatever the time of the month.

Everyone's welcome

Whatever your gender, if you have a menstrual cycle, then this book is for you – period!

Not every woman menstruates, but not everyone who menstruates identifies as a woman either. As such, we've aimed to use inclusive language throughout this book so that, however you identify, you know that you have the right to support, advice and information, which will help you better understand your body so that you can work with your menstrual cycle.

If you don't identify as a woman but do have periods, your menstrual cycle may pose additional challenges and complexities. For example, if your periods heighten feelings of gender dysphoria then there is specific support you might seek to help. For more information on this, turn to page 49.

Biology basics

Let's begin with a quick whistle-stop tour of the reproductive structures relating to the menstrual cycle to get you acquainted with your body and why your period occurs.

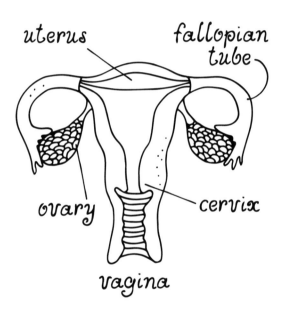

Typically, each month an egg is released from one of your ovaries (alternating sides every month), and travels down your fallopian tube towards your uterus (womb). While this is happening, the inner lining of your uterus (the endometrium) is preparing for potential pregnancy by getting thicker and spongier. In a textbook scenario, if the egg is fertilized on its travels, it will implant itself into the inner layer of the uterus so it can develop into an embryo. If the egg remains unfertilized, it will continue its journey to the uterus but will degenerate and fall away along with the blood-rich uterus lining. This is released from your vagina as your period.

Period number crunching

- The average age for your first period (menarche) is 12.4 years.

- The average person will have around 480 periods in their lifetime (fewer if you have any pregnancies).

- Your periods will usually stop between the ages of 45 and 55 (menopause). The average age for this to happen is 51.

Your first period

Whichever emotions were attached to it – be it excitement, joy, worry or shame – your first period (known as "menarche") is a key step towards maturity.

According to research, menarche usually occurs between the ages of ten and 16.

Think back to your first period. How did it make you feel? How was it received by others? If you still struggle with negative feelings today, such as shame, ask yourself if that stems from those early years and the messages you received about your period. We'll take a deeper dive into the shame surrounding periods in Chapter Three (from page 140), but right now, take a moment to think about your personal experience of menarche. If, back then, others made you

feel that your periods were shameful, it reflects more about them than you. You can break away from that mindset. Your menstrual cycle can be a wonderful thing once you learn to work with it. However you were made to feel when you were young, it's time to start celebrating your amazing, beautiful body right now!

Journalling about your early experiences can be extremely cathartic. Use the prompts over the following pages to write down your own thoughts, feelings and experiences surrounding your first period. Stay open and honest, and jot down whatever comes to mind – you might unlock the door to some deeply held beliefs about your period and your body...

Period journal prompts

Use these pages to jot down and unravel any thoughts that arise from the prompts. Remember, these notes are private – no one has to see them but you!

How old were you when you got your first period? Where were you? How did you feel?

_____

_____

_____

_____

Who did you tell and how did they react?

_____

_____

_____

_____

Were your periods framed positively or negatively?

_____

_____

_____

What messages were passed down from your female family members? Did the male figures in your life speak to you about periods?

_____

_____

_____

Compare your notes for the first question with how your family and friends reacted. Do you feel you've internalized other people's reactions?

_____

_____

_____

_____

The menstrual cycle

Your period is just one aspect of your menstrual cycle. This cycle comprises a series of hormone-driven processes that prepare your body for potential pregnancy, and it's divided into four distinct phases:

- Menstruation (your period)
- Follicular phase (before release of an egg)
- Ovulation (release of an egg)
- Luteal phase (after release of an egg)

On average, the entire menstrual cycle takes 28 days. In a textbook scenario, you'd get your period every 28 days. But really, who's average? In fact, a regular cycle is classed as completely normal if it lasts anywhere between 21 and 35 days (we'll take a look at irregular cycles on page 27), so if yours is a little shorter or longer than the 28-day cycle we're taught about at school, there's no need to worry.

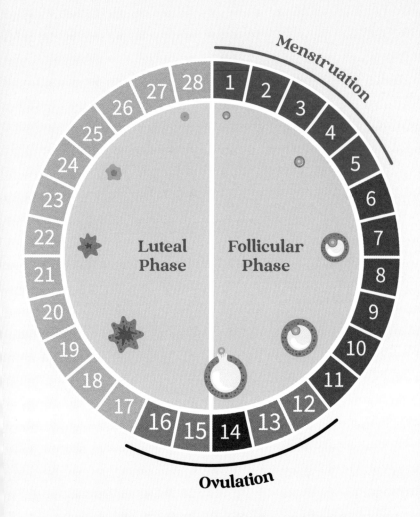

Over the following pages, we'll explore each phase, looking at what's going on with your hormones and your body, and how you might expect to feel both physically and emotionally during each phase of your cycle.

The follicular phase

The follicular phase starts on the first day of your menstrual cycle – so, day one of your period – meaning it overlaps with menstruation. It begins when the hypothalamus (the area of your brain that manages hormones) signals your pituitary gland to start releasing the follicle-stimulating hormone (FSH). This hormone signals your ovaries to start producing between five and 20 follicles, all of which contain an immature egg. The healthiest egg reaches maturity, while the rest are reabsorbed into your body. While this egg is maturing, it stimulates a surge in oestrogen, which stimulates the lining of your uterus to thicken.

The length of the follicular phase will differ from person to person and depends on how long it takes for your follicle to mature. Research suggests the average length of this phase is 16 days.

A long follicular phase

You'll know you have a long follicular phase if your cycle is long overall (between 29 and 35 days). If this sounds familiar, it's likely it's completely normal for you, although low vitamin D levels or long-term birth control can also lengthen your follicular phase. A longer follicular phase is unlikely to affect fertility.

A short follicular phase

If you have a shorter menstrual cycle (21–27 days), it's likely you have a shorter follicular phase. Again, this is probably completely normal for you, though research suggests factors that might shorten the follicular phase include heavy caffeine and alcohol intake. It also tends to shorten as you get closer to the menopause.

A short follicular phase may lessen your chance of conceiving. If you're struggling to conceive, visit your doctor, who will be able to work with you to determine the cause. Remember, a shorter cycle doesn't always equal difficulty conceiving, so don't panic if this is you!

The follicular phase – feelings and emotions

Your entire menstrual cycle is governed by hormones, which means the way you feel throughout is going to shift and change as these hormones rise and fall.

Because the follicular phase overlaps with menstruation at the start, you'll likely initially experience menstrual phase symptoms *and* premenstrual syndrome (PMS – which you can read more about from page 38). However, as both your oestrogen and FSH hormone levels begin to rise towards the middle and latter end of the follicular phase, this can bring with it some welcome and positive physical and emotional changes, including:

- Increased energy levels
- Improved mood
- Increased feelings of sociability
- Clearer skin
- Higher sex drive

Ovulation

As your oestrogen levels rise towards the latter end of the follicular phase, your pituitary gland is stimulated to release luteinizing hormone (LH) – the hormone responsible for ovulation.

Ovulation refers to the release of a mature egg from one of your ovaries (you have two, and they take it in turns to ovulate). In a textbook 28-day cycle, ovulation will occur in the middle, around day 13 or 14, and in the 12 to 24 hours following ovulation, while the egg is travelling along your fallopian tube, it's the time you're most likely to become pregnant if you have sex and your egg meets with sperm.

Signs you're ovulating include:

▶ A slight rise in basal body temperature (typically less than 0.3°C)
▶ Your vaginal discharge becomes clear and slippery, like egg white (more on discharge on page 32)
▶ You might experience ovulation pain (see page 23)

Ovulation – feelings and emotions

With your oestrogen levels at an all-time high during ovulation – and with your testosterone levels peaking – many people with menstrual cycles report feeling not only at their happiest and most energized during this phase, but also their sexiest and most desirable (yay ovulation!). Research shows there may be a biological driver behind these feelings and emotions – after all, if you feel friskier and more attractive, you might be more likely to have sex, which could result in pregnancy. (Note: if you want to have sex with someone who has a penis but don't want to become pregnant, always use contraception, whatever phase of your cycle you're in!)

It's not all good news, though: some research shows that, due to fluctuations in oestrogen, LH and progesterone around the time of ovulation, you might also experience mood swings similar to the ones experienced during PMS. So if you don't experience the textbook "happy phase", give yourself a break – it's "normal" and your feelings are valid.

Ovulation pain

Physical pain and cramping is most often associated with your period, but if you experience pain during ovulation as well, you're not alone and it's quite common – though less talked about.

Each month during ovulation, an egg bursts out of its follicle, and some people feel this follicular rupture as a sharp pain in their lower abdomen. It's usually mild, but can be quite uncomfortable, and normally lasts between three and 12 hours while you're ovulating.

Because you ovulate on alternate sides each month, you'll likely notice the pain shifts sides each month – or you might notice the pain is worse on one side, or even only feel it on one side.

Ovulation pain is usually nothing to feel concerned about. However, if yours is severe, lasts longer than three days and/or is accompanied by unusual symptoms such as a heavy bleed, it's important to visit your doctor as it could be an indication of an underlying condition, such as endometriosis (more on page 52).

The luteal phase

The luteal phase denotes the second half of your menstrual cycle and starts when the newly released egg begins to travel along the fallopian tube. During this phase, your level of the hormone progesterone rises, which helps the lining of the uterus thicken in preparation for pregnancy.

If you become pregnant during this cycle, the fertilized egg will implant in the lining of the uterus. If you don't become pregnant, your progesterone level will drop back down and the lining of the uterus will shed (your period).

The luteal phase can last between 11–17 days, though for the majority of people it lasts 12–14 days. A short luteal phase is considered to be 10 days or less and can make conception trickier, as the lining of the uterus doesn't have as long to develop; whereas a long luteal cycle might indicate an underlying condition, such as polycystic ovary syndrome (PCOS – more info on this on page 54).

Unlike the follicular phase, the luteal phase of your cycle tends to remain constant as you age.

The luteal phase – feelings and emotions

Because of the fluctuating levels of oestrogen and progesterone in your body during the luteal phase (around day 20 of a textbook cycle, they take a dramatic nosedive), you can be left feeling low, sensitive and anxious around this time. Common mental, emotional and physical symptoms of the latter luteal phase include:

- A dip in energy levels
- Increased feelings of anxiety
- Feeling low
- Becoming more self-critical
- Tender breasts
- Feeling bloated
- Skin problems, especially breakouts

Sounds fun, right? The good news is, by starting to track your cycle and beginning to understand why you feel the way you do, you can start to implement greater self-care during this time to help you cope (more on this in Chapter Two, from page 72).

Menstruation (your period)

Following the luteal phase, and if pregnancy doesn't take place, your hormone levels drop to the point when they can no longer sustain the built-up lining of your uterus and it sheds away, leaving your body as your period.

The first day of your period is classed as day one of your menstrual cycle (making your cycle easy to track, as you always know when to start counting). Bleeding tends to last between two and seven days, and during every period you'll lose 5–80 ml of blood (that's around one to six tablespoons).

The follicular phase starts again on the first day of menstruation, and the whole cycle begins again.

Irregular cycles

Many people who have periods have a regular cycle, even if it's a little shorter or longer than the average 28 days we're taught about at school. However, many people do not. In fact, some 14–25 per cent of people who menstruate have irregular periods. Your periods might be classed as irregular if:

- The length of your menstrual cycle keeps changing
- The gap between your periods is less than 21 days or more than 35 days
- The amount you bleed during each period fluctuates from one month to the next

An irregular cycle is not usually a sign that anything's amiss and could simply be normal for you. However, it can be worth visiting your doctor, as there may be an underlying cause, such as PCOS (see page 54). We'll take a closer look at the reasons behind irregular cycles and changes to your cycle from page 59.

Light or heavy?

You might have heard people talk about light or heavy flow in relation to periods, but have you ever wondered what exactly constitutes a light or heavy flow – and how yours measures up? Well, a review updated in 2023 suggests the following:

- **Light flow:** bleeding less than 5 ml during your period. This might equal roughly one tampon's worth of blood.
- **Normal flow:** bleeding between 5–80 ml during your period. This might equal around two to 16 tampons' worth of blood.
- **Heavy flow:** bleeding more than 80 ml during your period. You might find you need to change your tampon every one to two hours at certain points while you're bleeding, or you may need to use two different types of period products (such as a tampon and period underwear) at the same time.

Your flow may differ throughout your lifetime (or even from month to month) depending on certain factors, such as your age (periods tend to get shorter but heavier as you reach perimenopause), your diet and exercise regime (exercising frequently or eating too little can make your periods lighter or make them stop), or having an underlying condition like endometriosis or PCOS (both of which can affect the heaviness of your flow).

If you're concerned that your periods are especially light or heavy, it's worth speaking to your doctor, who will be able to rule out any underlying health concerns.

Why does the colour change?

While period blood is the same as the blood that flows through your veins, it also comprises tissue from the lining of your uterus. This means that it has a different consistency and – sometimes – colour to the blood that you might see flowing from a wound in your skin.

The colour of the blood can even change during the course of your period, too.

Pink blood

This is common at the start of your period, when your flow might be light. The blood can appear pink because it's mixed with your usual vaginal discharge.

Bright red blood

This is quite usual when your period is at its heaviest and is a sign that you have a steady flow of fresh blood. The bright colour of the blood means it's not been in the uterus for very long, instead passing quite quickly from the uterus.

Dark red/brown/black blood

A darker red colour, or even brown or black blood, is simply a sign the blood has been lining your uterus for a longer time. It's quite usual for the colour of your period to darken as your flow progresses, as older blood, from deeper in the uterus lining, is shed towards the end of your period.

While these are the most common colours you might see during a regular period, there are sometimes other colours you might notice, too, which could indicate an underlying problem, such as an infection.

Orange, grey or green tinges

If you notice orange, grey or green streaks in your period blood, or in your vaginal discharge, it could be a sign of infection, such as bacterial vaginosis (BV) or a sexually transmitted infection (STI). If you notice a colour that's not normal for you, seeking advice and support from your doctor is the quickest way to ensure everything is fine or, if it's not, to get the correct treatment as quickly as possible. Remember, an STI can happen to any sexually active person and is nothing to be ashamed of, so please don't feel embarrassed. Seeking treatment for it is no different to seeking treatment for any other illness – your doctor is there to help, not judge.

What's that wet patch?

In between your periods, you've probably noticed wet patches or dried white streaks in your underwear. No, it's nothing to be alarmed about; no, it's not dirty; and yes, it's perfectly normal. In fact, vaginal discharge is your body's way of keeping your cervix and vagina happy, healthy and free from infection.

Around the time of ovulation, this discharge also plays a vital part in keeping sperm alive to maximize your chances of pregnancy. (Note: if this is something you *don't* want to happen, always use birth control!)

The texture of your discharge will likely change depending on where you are in your menstrual cycle. It can go from being quite watery, to milky, to resembling raw egg white if you touch it (and don't be afraid of doing this, it's a completely normal bodily fluid – it will stretch between your fingers). This "egg white" discharge is a sign you're at your most fertile and happens around ovulation.

When should you worry?

If you notice your vaginal discharge smells "fishy", or if it's yellow, green, grey or streaked with blood, lumpy in texture, or if your vagina and/or vulva (the external parts of your sexual organs) become itchy, it's a sign something's up. There's no need to be embarrassed, but it is important to get it sorted. Take a trip to your doctor as it's possible you've picked up an infection, such as BV or an STI, or you may have thrush (an overgrowth of yeast). Your medical practitioner will be able to diagnose the cause and prescribe the best course of treatment.

Feeling hormonal?

As we've seen over the previous pages, your entire menstrual cycle (in fact, your entire *body*) is governed by your hormones. This means that, as your hormones fluctuate depending on where you are in your cycle at any given moment, your mood, energy levels and feelings of sociability and sexual desire can all fluctuate, too. Some people only attribute the more challenging moods and emotions to hormones, but actually, it's all of it!

In the next chapter, from page 72, we'll take a closer look at how you can work with your own unique cycle, which will help you harness those high-energy, sociable, happy moments, and also figure out some techniques to make the tougher moments more manageable, too.

For now, though, let's take a deeper dive into some of the more challenging aspects of your menstrual cycle, so we can understand a little better why we feel what we're feeling at certain times, and learn what's really behind those low, anxious, sometimes rage-filled moments.

Period pain

It's common to experience cramping in your lower abdomen, back and/or thighs around the time you get your period. In fact, the charity Women's Health Concern says around 80 per cent of people who have periods experience period pain (technical term: dysmenorrhoea) at some stage in their life. It's most common on day one of your period. This is because, right before you start bleeding, your oestrogen and progesterone levels start to fall which prompts your body to release prostaglandins, which cause your uterus to contract to release the blood (fun fact: if you have high levels of prostaglandins, the "contract" message can get sent to your bowels, too, which is why some people feel the need to poo right before their period starts).

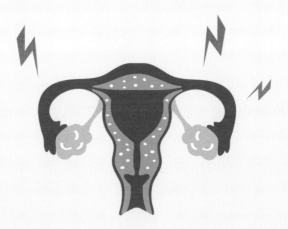

Coping with period pain

Period pain isn't fun, but usually it can be eased with some simple, at-home remedies…

- **Gentle movement:** a little light exercise, such as going for a walk, can be beneficial and help ease pain.
- **Rest:** don't feel like moving? Then rest up by taking a nap or curling up with a good book. We're all different, so do what works for you.
- **Apply heat:** holding a heat pad or hot water bottle to the site of the pain, or taking a warm, relaxing bath, has been scientifically proven in several randomized, controlled trials to help ease period pain.
- **Try a B-vitamin complex:** B vitamins – specifically thiamine (B1) and niacin (B3) – have been shown in some studies to help reduce period pain. Vitamin B6 has also been shown to help reduce low mood related to PMS, so take a B-vitamin complex to make sure you're completely covered!
- **Stop smoking:** if you're a smoker, this is another good reason to quit – studies show that smoking increases your risk of painful periods.
- **Take a painkiller:** an anti-inflammatory, such as ibuprofen, can inhibit the production of prostaglandins, which can ease period pain.

Mild period pain is pretty common, but if it's so bad that you can't get out of bed, or it's inhibiting your life in other ways, consult your doctor. Conditions such as endometriosis can make period pain excruciating, so don't suffer in silence.

Menstrual migraines

Menstrual-related headaches (MRH) can occur due to the decline in oestrogen in your body before your period starts. They're most common in the two days leading up to your period and/or during the first few days of bleeding, and according to the American Migraine Foundation, for nearly two out of three people who have periods and migraines, attacks occur around the same time as menstruation.

Menstrual-related migraines are rotten, but there are a few things you can do to help manage them. Eating small, regular snacks can help to keep your blood sugar levels stable, which can ward off an attack, as can ensuring a regular sleep habit at night, avoiding alcohol and sugary snacks, and limiting stress where possible (tricky in our always-on lives, we know). If you don't notice an improvement, visit your doctor, who may be able to prescribe medication to help.

What is PMS?

Premenstrual syndrome (PMS for short) is the name given to a collection of symptoms many people experience in the run-up to their period. Everyone's different, so your experience of PMS might not match your friends' or family members' experiences. Generally speaking, symptoms can hit any time after ovulation and usually last until a few days after your period begins. Some people might only have mild symptoms that last a few days, while others can experience PMS for a few weeks leading up to their period. A recent scientific review found that some 47.8 per cent of people who have periods experience PMS, and of these, 20 per cent have symptoms severe enough to affect their daily life.

PMS symptoms can be physical, mental and emotional, and may include:

- Abdominal pain and cramping
- Lower back pain
- Headaches
- Breast tenderness
- Nausea
- Constipation
- Low mood
- Anxiety
- Fatigue
- Anger

Breast tenderness

Cyclical breast pain (breast soreness linked to your menstrual cycle) is a common PMS symptom, often occurring in the week leading up to your period, and is thought to be down to hormones (of course!). When it comes to the research, though, the jury seems to be out as to the exact cause: some studies suggest it occurs because of low levels of progesterone, while others have found it could be down to an abnormality in the hormone prolactin. Making sure you're wearing a comfortable, supportive bra can help, as can avoiding caffeine and eating a low-fat, high-fibre diet.

Tracking your breast pain over a couple of months is a good idea (you'll be able to do that in Chapter Two), to ensure it's cyclical and therefore linked to your menstrual cycle. If, after tracking, you discover the pain is more persistent and seems unrelated to your cycle, it's important to make an appointment with your doctor, who will be able to check for other underlying causes.

Fatigue

Feeling wiped out before your period is one of the most common symptoms of PMS, and is thought to be due to a lack of serotonin, which can see your energy levels plummet. While mild to moderate fatigue is expected, if you're struggling to function it might be a sign of premenstrual dysphoric disorder (PMDD – more on page 44), in which case it's worth consulting a health expert.

Some of the best ways to help cope with PMS-related fatigue are to maintain a regular sleep pattern each night, drink plenty of water every day and avoid caffeine where possible. Regular, gentle exercise throughout your cycle also helps as it can boost your energy and strength. Most importantly, though, rest when you're tired (if you're able) – there's a lot going on in your body, after all, so giving yourself a break can really help.

Food cravings

Fluctuating hormone levels affect your neurotransmitters, which means you might start craving carbohydrates and sugary treats after ovulation right up to your period as your body does its best to raise your depleted energy and serotonin levels (chocolate, anyone?).

Listen, who are we to tell you what you should and shouldn't eat in the run-up to your period? After all, a little of what you fancy often does you good (in fact, chocolate also contains magnesium, which your body needs more of following ovulation). However, if you're struggling to get your cravings under control or you find you're binge eating, taking a few simple steps can help: aim to drink more water throughout your cycle, stock up on wholesome snacks so you have something healthy to reach for when cravings strike, opt for wholegrain over refined grains (such as brown rice instead of white), and avoid alcohol and caffeine, which can make everything seem worse!

Irritability, low mood and downright rage

It's the stereotypical PMS symptom – the one many people who menstruate feel defined by and are, sadly, sometimes mocked for ("time of the month, is it?").

We saw on page 40 that low serotonin levels can be the cause of low energy in the luteal phase of your cycle, and this same low serotonin is also one of the reasons for the irritability, sadness, low mood, anger and downright rage you might feel at this time. We say "one of the reasons" because other causes of rage include the aforementioned comments from less-than-helpful, insensitive individuals, and in these instances your anger is completely justified!

Tracking your mood (more on this in Chapter Two) is a good way of figuring out your unique emotional patterns relating to your menstrual cycle and, once you have a clearer understanding of when you're likely to feel upset, angry or triggered, you can make small adjustments to your lifestyle in order to look after yourself (for example, if you know you

tend to feel low on days 26–28 of your cycle, you can plan ahead in future months by scheduling a duvet day rather than a big night out with friends). Other steps that can be helpful in managing the (sometimes) wildly fluctuating mood swings include taking regular gentle exercise, making sure you're eating a healthy balanced diet (the odd treat is totally fine) and taking a B-vitamin complex (B6 has been shown to help ease PMS-related mood swings). If you're really struggling, visit your doctor, who may advise medication or talking therapy. And of course, as previously mentioned, sometimes your rage is justified and can be a powerful sign that something in society is unfair, unjust or simply plain wrong. In which case, channelling your anger to help fight the patriarchal system that keeps us feeling oppressed might just be your best bet.

Premenstrual dysphoric disorder

If your PMS symptoms are severe, you might have premenstrual dysphoric disorder (PMDD). As with PMS, PMDD will occur during the luteal phase of your cycle, between ovulation and the start of your period. As well as symptoms typical of PMS (see page 38), if you have PMDD you may also experience additional symptoms that can seriously affect your life. These can include:

- Feelings of hopelessness
- Extreme sadness, anxiety and/or rage
- Extreme fatigue
- Feeling totally overwhelmed
- Trouble concentrating
- Lack of interest in hobbies and activities you usually find enjoyable
- Suicidal thoughts or feelings

While PMDD is classed as an endocrine disorder (meaning it's hormone related), it's also classed as an official mental health problem in the DSM-5 (the manual doctors use to classify and diagnose mental health disorders) because of the severity of the symptoms.

It's thought that PMDD might be caused by genetic variations, although other studies have found it could be linked with past trauma and/or high stress levels. Being a smoker might also impact your hormone sensitivity, which could increase the severity of your symptoms.

If your PMS symptoms are severe, negatively impacting and interrupting your life each month, or if they're making you feel suicidal or that life isn't worth living, it's important to make an appointment with your doctor so they can offer you impartial and confidential support and advice.

 Getting professional support with PMS and PMDD symptoms

Sadly, because of learned shame, many of us don't feel comfortable discussing periods even though they're a normal part of life. This means that even though asking for help with PMS/PMDD sounds simple enough, in reality it can sometimes feel deeply uncomfortable.

We'll take a closer look at where this shame comes from and how to begin dismantling it in Chapter Three. For now, though, if you need to make a medical appointment, the following might help you:

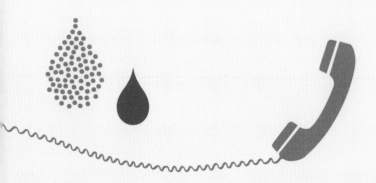

- Make your appointment request online, rather than in person or over the phone. This can often feel less threatening.
- Request an appointment with a female doctor, or one who specializes in gynaecology or mental health.
- Keep a record of your symptoms and cycle before your appointment. Your doctor will likely ask you to do this, so having the information to hand means you're one step ahead!
- Consider asking a trusted relative or friend to attend your appointment with you, so they can advocate for you if you're struggling with your mental health right now.
- Write down any questions you might have before the appointment, so you don't forget anything.

Premenstrual exacerbation

If you have an existing mental health condition, you might find your symptoms get worse during the luteal phase of your menstrual cycle. According to the International Association for Premenstrual Disorders, the psychiatric disorders that can be aggravated by your menstrual cycle include:

- Alcoholism
- Anxiety disorders
- Bipolar disorder
- Depression
- Eating disorders
- Schizophrenia

If you have been diagnosed with a mental health disorder, or suspect you have a mental health disorder (in which case it's important to visit your doctor for support and guidance), tracking your menstrual cycle can be an important tool for monitoring the severity of your symptoms. This will help you identify when you might need additional support from friends, relatives or professionals.

Coping with periods if you're trans or non-binary

If you're transgender or non-binary, you may experience heightened gender dysphoria around the time of your period, as well as the more common PMS symptoms. In fact, a 2022 study that looked into transgender and non-binary adolescent experiences of menstruation showed that some 93 per cent of participants reported experiencing menstrual-related dysphoria. This can often be exacerbated by societal views surrounding periods and the language used in relation to period products.

Gender dysphoria can bring with it feelings of depression and shame, and even suicidal ideation. If you're struggling, please know you have nothing to feel ashamed of and that you are not alone. There are people out there you can talk to at LGBTQ+ charities and counselling services who can offer ideas on how to ease dysphoria. Using period products designed for trans men can also be a huge help, such as period boxers. Make an appointment with your doctor to discuss the support options available to you.

Conditions that can affect your cycle

Certain uterine and ovarian disorders can affect both your hormones and your menstrual cycle, which sadly makes you more likely to experience an irregular cycle, painful periods and/or heavy bleeding during your period. The most common of these conditions include:

- Uterine fibroids
- Endometriosis
- Polycystic ovary syndrome (PCOS)
- Ovarian cysts
- Primary ovarian insufficiency

If your periods are painful, heavy or irregular, you might well be unknowingly suffering from one of the above disorders. Over the following pages, we'll take a closer look at each, to help you learn to identify the symptoms and understand why you might be struggling with your periods.

Uterine fibroids

Uterine fibroids are non-cancerous growths that develop in or around the uterus. They are common: two in three people who menstruate will develop at least one fibroid during their lifetime, and they occur most commonly between the ages of 30 and 50. They can vary in size considerably: some might be the size of a pea, while others can be the size of a melon. If your fibroids are small, you might be unaware you even have them as you might not have any symptoms, and are typically benign and therefore harmless. If you do have symptoms, though (and one in three people do), these might include:

- Heavy and/or painful periods
- Abdominal and/or lower back pain
- The need to urinate more frequently than usual
- Constipation
- Pain or discomfort during sex

See your doctor if you're experiencing any of the above symptoms, as treatment is available, including medicines that can shrink the fibroids.

Endometriosis

Endometriosis is a condition where tissue similar to the lining of the uterus develops in other parts of the body, such as the ovaries or fallopian tubes. Each month with your menstrual cycle, this tissue reacts in the same way as the lining of the uterus, building up before breaking down and bleeding. However, unlike the tissue inside your uterus, which comes away as your period, this blood has nowhere to go. The result of this is pain, inflammation and the formation of scar tissue.

Endometriosis can have a number of both physical and emotional symptoms, including:

- Abdominal and/or lower back pain (usually worse during your period)
- Heavy periods
- Extreme period pain
- Pain during and/or after sex
- Pain when going to the toilet during your period
- Nausea during your period
- Constipation during your period
- Blood in your urine and/or faeces during your period

- Difficulty getting pregnant/infertility
- Depression

If you have symptoms of endometriosis, it's important to see your doctor. While there is currently no cure, treatment can help to manage and ease your symptoms. There are a number of options available to you, from pain management to hormone treatment and surgery.

If endometriosis is negatively affecting your life, you don't need to suffer in silence. There are support groups who understand what you're going through and can provide vital emotional support (see our resources on page 156, or speak with your doctor who will be able to direct you to relevant services).

Polycystic ovary syndrome (PCOS)

Because many people who have polycystic ovary syndrome (PCOS) don't have any symptoms, it's hard to know how many people are living with the condition. However, it's thought to be a common condition, with estimates suggesting it affects one in 10 people who menstruate. If you do have symptoms, these will usually appear during your late teens and early twenties, and can include:

- Irregular or absent periods
- Excessive hair growth (hirsutism), usually on the face, chest, back or buttocks, due to an increase in androgen hormones
- Weight gain
- Thinning hair or hair loss
- Oily skin and/or acne

- Irregular or absent ovulation (signalled by irregular or absent periods)
- Difficulty getting pregnant/infertility

Infertility is one of the most common symptoms of PCOS and is actually an indicator of the condition; many people don't realize they have it until they visit their doctor to discuss issues with fertility.

If you're struggling with any of the above symptoms, it's important to visit your doctor for support and diagnosis. While there's currently no cure for PCOS, there are things you can do and treatments available to help manage symptoms, including lifestyle changes (such as weight loss if you're currently overweight), medicines and IVF treatment.

Ovarian cysts

An ovarian cyst is a fluid-filled sac that develops on your ovary, which is common and usually harmless. Its symptoms usually present when the cyst bursts, or if it's large and causes the ovary to twist (ovarian torsion). These symptoms can include:

- Heavy and/or irregular periods, or very light periods
- Pelvic pain (which can be sudden and sharp)
- Constipation or pain while going to the toilet
- Pain during sex
- Bloating or feeling full
- Difficulty getting pregnant

Ovarian cysts can be diagnosed via ultrasound. Often, cysts disappear on their own within a couple of months; a second ultrasound can confirm whether it's cleared up. If treatment is needed, this can be in the form of a laparoscopy (a type of keyhole surgery).

Primary ovarian insufficiency (POI)

This is the term used to describe when a person's ovaries stop working normally before the age of 40 (the average age of menopause is 51, so it's significantly earlier). This could be due to the fact the person has fewer eggs than usual, or it might mean the ovaries are not maturing or releasing eggs correctly. It differs from menopause, which is triggered by your ovaries naturally running out of eggs.

With POI, there may still be eggs inside the ovaries, and the person might still have a menstrual cycle (although it might be irregular). Symptoms include:

- Irregular periods
- Decreased sex drive
- Feelings of irritability
- Discomfort during sex
- Night sweats and/or hot flushes
- Difficulty getting pregnant

It's usually treated with hormone therapy, so visit your doctor if you're experiencing symptoms for a consultation.

Vaginal infections

Your vagina is a naturally acidic environment. However, when your period arrives, the higher (more alkaline) pH level of menstrual blood can cause your vaginal pH levels to become imbalanced. This can lead to a higher risk of developing an infection around the time of your period, such as thrush, due to the ability of yeast to thrive in higher pH levels.

Common symptoms of thrush include a thick, white discharge (resembling cottage cheese), extreme itching, a burning sensation while peeing, and pain and/or itching during sex. Thrush can be easily treated with an over-the-counter antifungal treatment, in the form of an oral pill or vaginal pessary, but if it's the first time you've experienced symptoms, it's worth a visit to your doctor just to rule out anything more untoward.

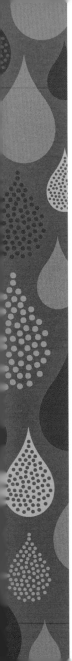

Changes to your menstrual cycle

As we've seen on page 27, many people have an irregular menstrual cycle. Often, this is nothing to worry about and is simply normal for you. However, if your menstrual cycle has suddenly changed – perhaps it has become more irregular, you've skipped a period or your periods have stopped altogether – there is likely to be an underlying cause.

A missed period is, of course, one of the first signs of pregnancy, so if your periods have stopped, it's important to take a pregnancy test. If the test is negative, or you know there's no way you can be pregnant, there are many other reasons your period might suddenly have become irregular or stopped. We'll look at some of the most common causes in more detail over the next few pages.

Sudden weight change

A significant amount of weight loss or gain can impact the regularity of your periods. If this is intentional, tracking your menstrual cycle alongside your bodily changes will help you see if this is the underlying cause; however, if the weight loss or gain is unexpected and not caused by dietary or lifestyle changes, please consult a doctor.

Weight gain

An increase in fat stores can lead to a hormonal imbalance. This is because adipose tissue (fat) can trigger an increase in oestrogen production and hinder ovulation. Weight gain can also lead to heavier periods, because obesity-related inflammation has been shown to delay endometrial repair and increase blood loss.

Weight loss

Sudden or rapid weight loss can also interrupt your periods, making them lighter and irregular, or causing them to stop altogether. You need at least 22 per cent body fat in order to menstruate regularly. Dipping below this can cause a stress response that alters your hormone levels, which not only affects your cycle and fertility, but can also be harmful to your bone health.

Aim to eat a healthy, balanced diet, packed full of protein, carbohydrate, healthy fats and fruit and veg to help you maintain a healthy weight, which will in turn help to regulate your menstrual cycle.

Overexercising

Heavily linked to period irregularities due to weight loss, excessive exercise can also cause your periods to become light, irregular or stop altogether. This is thought to be because rigorous exercise and training places your body under extreme stress and, coupled with extreme energy expenditure and inevitable weight loss, your body will stop ovulating as a means of conserving energy.

Lack of periods can be common in people who have to follow strict training programmes, such as athletes and dancers. However, it can be highly detrimental to your health as it can lead to lower bone density.

If you do lots of exercise, fitting in adequate rest days and eating well, to ensure you're not in a calorie deficit is vital.

You can find more advice about what to do if your periods have stopped altogether (known as amenorrhea) on page 64.

Chronic stress

As we've seen over the previous pages, weight loss and rigorous exercise – which place your body under nutritional and physiological stress – can both cause your periods to become irregular or stop completely. However, did you know that emotional stress can also play havoc with your menstrual cycle?

The increase in cortisol production when you're experiencing chronic stress can be enough to mess with your hormones and interrupt your periods. If you're experiencing high stress levels and your period is late, or you skip your period completely, it's a sign you may need to take some measures to lower your cortisol levels. We'll look into self-care for period health in the next chapter (from page 104), but the basics include eating a balanced, healthy diet, ensuring you get adequate rest, addressing any issues causing you worry or anxiety, and establishing a healthy sleep regime.

Iron deficiency

Low iron levels can disrupt your menstrual cycle, with evidence suggesting that severe iron deficiency can cause you to have shorter, lighter periods in a bid to preserve your iron stores. It can also cause you to stop menstruating altogether.

However, it's important to note that for people who menstruate, heavy periods are the leading cause of iron deficiency. This can therefore result in something of a vicious – and irregular – cycle: after several heavy periods in a row, your body might then struggle to replenish iron levels, causing your next period to be late, or skipped completely.

If you're experiencing period irregularity as well as other symptoms of iron deficiency, which include extreme fatigue, shortness of breath, a rapid heartbeat and generally feeling weak, it's important to see your doctor. Iron supplements should help, but seek medical advice first.

What is amenorrhea?

If you've missed three periods in a row and you aren't pregnant, it's known as amenorrhea (specifically, secondary amenorrhea – primary amenorrhea is when your periods haven't started by the time you're 15).

We've covered the main lifestyle causes of amenorrhea over the previous pages, which include sudden weight change, overexercising and stress. However, there might be a hormonal reason behind your stopped periods, such as polycystic ovary syndrome (see page 54), a thyroid problem or premature menopause. Some medicines can also cause amenorrhea, including certain antidepressants and allergy medications.

If your periods have stopped, it's important to visit your doctor to determine the cause. They will probably ask you to take a pregnancy test first to rule this out, and will likely ask you about your family history, your periods before they stopped, and maybe run tests to help determine a diagnosis and suggest the most suitable plan of action.

Perimenopause and irregular periods

Perimenopause refers to the period of time that precedes menopause (the point at which you haven't had a period for a year). Perimenopause can last over ten years and, with the menopause beginning on average between the ages of 45 and 55, this means you might start getting perimenopausal symptoms as early as your thirties.

Often (though not always), one of the first signs of perimenopause is irregular periods, which might also become heavier but shorter too. If your periods have become irregular and you're experiencing other perimenopause symptoms, such as mood swings, heightened anxiety, "brain fog", hot flushes, heart palpitations, night sweats, vaginal dryness or difficulty sleeping, your doctor might be able to help. Receiving support early can reduce the impact perimenopause can have on your life. Treatment can include lifestyle changes and/or hormone replacement therapy (HRT) to minimize the effects of your body's natural hormone changes. In recent years, more people are speaking out about perimenopause – there are so many great resources on this topic – so know you're not alone and don't be ashamed to seek support if you need it.

Periods and contraception

If you use a hormonal contraceptive, such as the 21-day combined pill, vaginal ring or birth control patch, it's worth noting that during your break week (when you stop taking/using the contraception), you aren't having a regular period. Instead, when you bleed during this time it's known as a withdrawal bleed, because of the drop in hormone levels in your body. While similar to a menstrual period, a withdrawal bleed will likely be lighter, shorter and bring with it fewer PMS symptoms. It's also worth noting that sometimes your period can stop altogether.

Other forms of contraceptive, such as the copper intrauterine device (IUD) and progesterone-only contraception (which comes in the form of tablets, injections and the implant), can also affect your periods. The copper IUD, for example, potentially makes your periods heavier (in fact, research has shown that this form of contraceptive can increase menstrual blood loss by around 50 per cent).

Hormonal contraceptives are commonly prescribed when people suffer with bad PMS. However, do be aware that while they can lessen symptoms, they are merely masking the problem rather than addressing any underlying cause.

If you're considering taking a hormonal contraceptive to help deal with heavy periods, it might also be worth asking your doctor to investigate further, in case there's another issue at play such as endometriosis or fibroids.

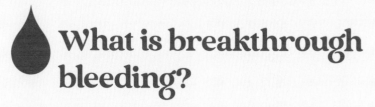

What is breakthrough bleeding?

Breakthrough bleeding refers to any bleeding or spotting that occurs when your period isn't due. There are a few reasons why it might happen:

- You've started taking a new hormonal contraceptive. It's not known why it happens, but some doctors believe it's down to your body adjusting to the hormones. If you forget to take a pill, you might also experience breakthrough bleeding.
- You have an STI or other infection, such as chlamydia.
- You have an underlying medical condition, such as endometriosis or fibroids.
- You have a sensitive cervix which has become irritated, for example, after sex.

If you experience breakthrough bleeding, it's important to see your doctor, who can determine the underlying cause.

Period products: what to use?

When it comes to period products, there are so many options to choose from it can become a bit bamboozling!

If you've been having periods for a while, it might be that you still use the period products you started out with back in your early teens, simply because it's all you've ever known. It's worth taking a little time to investigate other options though, as there might now be period products on the market that are easier, more convenient and better suited to your lifestyle.

While single-use disposable period products, such as pads and tampons, are still most widely used, reusable products, such as menstrual cups, reusable pads and period underwear, offer a more environmentally friendly choice and can be super convenient (for more information, see page 70).

Tampons and toxic shock syndrome

Tampons and menstrual cups are associated with toxic shock syndrome (TSS) – a very rare but life-threatening infection. It's caused by bacteria, including Staphylococcus aureus. Tampons can hold onto these bacteria and then, when removed (especially if not fully saturated), can cause small tears in the vagina, allowing the bacteria to enter the body.

Symptoms include a high temperature, muscle aches, a raised skin rash and flu-like symptoms. If you think you might have TSS, ask someone to take you to seek medical attention immediately, and tell the doctor your symptoms and that you've been using tampons.

When using tampons or a menstrual cup, good hygiene is key. Read the guidelines on the pack and change your tampon (or empty the cup) accordingly. For tampons, this is usually every four to eight hours; for menstrual cups, it's usually every ten to 12 hours. You should also always use the right-absorbency tampon, too – never higher than you need. Typically branded and non-branded tampons are purple for light flows, yellow for regular flows, green for medium-heavy flows, and orange for very heavy flows.

Sustainable period products

If you use single-use period products simply because it's what you've always used, perhaps now is the time to consider having a more environmentally friendly period?

According to leading environmental organization Friends of the Earth, conventional disposable pads contain roughly 90 per cent plastic – in fact, one pack of disposable period pads can be the equivalent of four plastic bags. In a world where so many of us are now consciously trying to reduce our plastic waste, period products can be an overlooked but vital issue to address.

Many people use single-use period products for convenience, without realizing reusables are just as easy – if not more so! With a little practice, menstrual cups can be easier to use than a conventional tampon, while period underwear now comes in an array of stylish and sexy designs. If cared for correctly, they can last for years:

simply remove your period pants and soak them in cold water after use, then add them to your regular wash on a delicate cycle – voila!

If you're not ready to swap to fully reusable period products, why not consider switching to an eco-friendlier single-use choice? You can now buy biodegradable and plastic-free pads and tampons, which are good for your body and kinder to the planet, too.

Chapter Two: Your Cycle and You

In Chapter One, we learned the basics: understanding the menstrual cycle and the issues that can affect it. Chapter Two is all about *you*: your cycle, your periods, your body and your emotional health and well-being. You'll learn about the importance of tracking your cycle, including how this one simple act can help you better understand your fluctuating moods and premenstrual symptoms (plus how to prepare for, manage and work with them); while journalling exercises will allow you to explore your relationship with your body. Finally, we'll take a look at the lifestyle changes you can implement to help you regulate your cycle and work with your body. Ready to feel empowered? Let's take a deep dive together...

What is period tracking?

In its simplest form, period tracking involves making a note of the dates your period starts and ends every month. By doing this over the course of several months, you'll begin to notice a pattern in your menstrual cycle and period length, and will be able to more accurately predict when your next period will start.

By keeping a note of additional information during the course of each cycle – such as how heavy your bleeding was, any cramping you experienced, how you were feeling emotionally, your energy levels, any food cravings you experienced, skin changes, etc. – you can begin to build up a bigger picture of your unique menstrual cycle.

Physical benefits of period tracking

Tracking your period is a great way of learning what's normal for *you*. This is not only important for your own peace of mind, but can also help you notice if things go off track, such as breakthrough bleeding, a particularly heavy period or worsening PMS. Tracking your period is also a handy way of keeping tabs on your fertility, helping you work out when you'll be most fertile if you're trying for a baby or least fertile to avoid pregnancy (although it's recommended to always use contraception whatever point you're at in your cycle for this).

Finally, if you need to visit your doctor about an issue related to your menstrual health or fertility, they will most likely always ask you when your last cycle began and what other symptoms you're experiencing. By tracking your cycle, you'll have this information to hand, which can speed up potential referrals or prescriptions. From page 80, you'll find a fill-in annual period tracker, as well as three months' worth of monthly trackers, to help you log all the information you'll need to start your tracking habit.

Emotional benefits of period tracking

Getting acquainted with your menstrual cycle has numerous benefits. Emotionally speaking, tracking your cycle can be an empowering act: after all, knowledge is power, and the more you understand your own mind and body – including cyclical rhythms, moods and emotions, cravings and fluctuating energy levels – the more you can begin to use this information to your advantage. For example, once you're familiar with your monthly cycle and how it affects your moods and energy, you can diarize key social and work events when you know you'll be feeling at your most bubbly, plan in the toughest workouts of your fitness regime while your energy is peaking, and block out time for self-care when you know you'll likely be feeling more drained and/or anxious.

How to track your cycle

To start tracking just your period, you don't need anything more than a diary or calendar and a pen: simply note down the start and end date of your period with an asterisk. To glean a little more data, you could use a journal to note down what your flow is like on each day of your period.

Tracking your entire cycle simply takes this a step further: as well as making notes during your period, you continue the habit for the entire length of your cycle (i.e. a whole month). You'll be thinking about your mood, energy levels, appetite, sex drive etc. Some days, you might only feel like writing a word or two (e.g. "excitable"; "bloated"), while on others you might feel inspired to really get into all your feelings, emotions and triggers that day.

There are lots of apps to help you track your cycle too, and you can also upload data to smartwatches, but journalling with pen and paper can feel more personal and allows you to write as little or as much as you'd like.

Your period tracker

Over the following eight pages, you'll find a series of calendars designed to help you start tracking your cycle.

The first one, on pages 80–81, is a simple annual cycle tracker, listing each month of the year. Simply write the year at the top and create your own symptoms key (either colour coded or by choosing different symbols), then fill in which days you have your period, along with any days you experience PMS symptoms. There's also space below the tracker where you can note your period and cycle length each month, along with space for additional notes where you can add any relevant patterns or observations you notice, for example, whether you had a particularly heavy period one month.

From pages 82–87, you'll find three monthly calendars. You can use these to track each month of your cycle in more detail. Write the name of the month at the top of the page and add the dates, then jot down in each square how you feel that day. Think about physical and emotional details – write down anything that feels pertinent to you. For example: "period starts"; "heavy bleeding"; "cried a lot"; "chocolate cravings"; "happy"; "anxious"; "sociable"; "turned on". If it felt notable, jot it down!

There's space for three months' worth of cycle tracking because if you notice anything that might warrant a visit to your doctor, they'll often ask you to track your cycle for two to three months to help them spot patterns that can help point them towards a correct diagnosis. It's not just for the medical professionals, though! Over the course of several cycles, you'll be able to start to figure out your personal trends and fluctuations – the times you feel great and times you feel low. This is important in helping you feel more empowered and better able to understand why you feel the way you do.

Year: Your annual

	1	2	3	4	5	6	7	8	9	10	11	12	13	14
Jan														
Feb														
March														
April														
May														
June														
July														
Aug														
Sept														
Oct														
Nov														
Dec														

Symptoms key

Light	
Medium	
Heavy	
Mood swings	
Headache	
Cramps	

period tracker

15	16	17	18	19	20	21	22	23	24	25	26	27	28	29	30	31

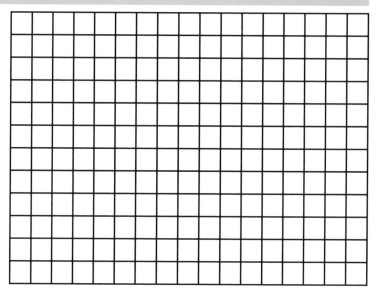

Cycle/period length

Jan	/	July	/
Feb	/	Aug	/
March	/	Sept	/
April	/	Oct	/
May	/	Nov	/
June	/	Dec	/

Month: **Your monthly**

Monday	Tuesday	Wednesday	Thursday
☐	☐	☐	☐
☐	☐	☐	☐
☐	☐	☐	☐
☐	☐	☐	☐
☐	☐	☐	☐

period tracker

Friday	Saturday	Sunday	Notes
☐	☐	☐	
☐	☐	☐	
☐	☐	☐	
☐	☐	☐	
☐	☐	☐	

Month: # Your monthly

Monday	Tuesday	Wednesday	Thursday
☐	☐	☐	☐
☐	☐	☐	☐
☐	☐	☐	☐
☐	☐	☐	☐
☐	☐	☐	☐

period tracker

	Friday	Saturday	Sunday	Notes
	☐	☐	☐	
	☐	☐	☐	
	☐	☐	☐	
	☐	☐	☐	
	☐	☐	☐	

Month: # Your monthly

Monday	Tuesday	Wednesday	Thursday

period tracker

	Friday	Saturday	Sunday	Notes
	☐	☐	☐	
	☐	☐	☐	
	☐	☐	☐	
	☐	☐	☐	
	☐	☐	☐	

You've tracked your cycle – now what?

Once you've tracked your cycle for a while, you'll be armed with a significant amount of data about your body, emotions and triggers, as well as the fluctuations you cycle through each month. You may have noticed your cycle is irregular, your periods last for more than seven days, or you've become concerned about aspects of your PMS – in which case, it might be an idea to visit your doctor. While this information alone can be useful, cycle tracking can also open up a world of personal self-discovery. This newfound knowledge can in turn lead to a deep sense of authenticity and empowerment as you begin to understand why you experience certain feelings, emotions and urges, and can start to align your life with your cycle where possible.

Over the following pages, you'll find some journalling questions that will help you dig deeper into the emotions, experiences and patterns you've noticed. You'll then be prompted to consider how you can use this knowledge to take greater control of your own life.

Period tracking prompts

Take a little time to think about everything you've discovered by tracking your cycle, then use the following prompts to unravel your thoughts and reflections.

On average, how long is your cycle?

___

___

Is your cycle regular?

___

___

What physical PMS symptoms do you experience (e.g. abdominal pain, tender breasts, etc.)?

___

___

Have you noticed any pattern as to when you begin to experience PMS?

___

___

How do you tend to feel emotionally in the run-up to your period?

_____

_____

_____

_____

When do you tend to feel energized and/or sociable during your cycle?

_____

_____

_____

_____

When do you tend to feel tired and/or less social during your cycle?

_____

_____

_____

_____

How do you feel about yourself when you have your period (e.g. sexy, strong, confident, powerful, small, weak, ashamed)? Where do you think these feelings come from? Do they tally up with any of the physical PMS symptoms you might experience?

_____

_____

_____

_____

_____

What small steps might you be able to take in order to align your life with your cycle? Think about your social activities, workout routine, and diet – are they working for or against you during your cycle?

_____

_____

_____

_____

_____

What habits, actions or things cause you most frustration around the time your period is due? What might you be able to let go of?

___

___

___

___

___

___

Has your relationship with your body changed while you've tracked your cycle?

___

___

___

___

___

___

How might you begin to honour your cycle? What rituals and self-care habits can you incorporate?

_____

_____

_____

_____

_____

_____

Has this tracking exercise helped you identify patterns in your moods and energy levels? How might this affect the way you approach your schedule and self-care needs?

_____

_____

_____

_____

_____

_____

Aligning your life with your cycle

One of the most empowering things about tracking and understanding your menstrual cycle is the ability to start aligning your life with its fluctuating rhythms. This can take you from a place of feeling out of control, overwhelmed and anxious, to a state of awareness and flow – a state in which you're not only able to predict and anticipate when you're going to feel a certain way, but are able to then work with those emotions and physiological states to improve your life.

Take a moment to look back over your monthly cycle tracker (pages 82–87), as well as your answers to your journalling prompts (pages 89–93). Can you spot any trends? Perhaps you've found that each month, you start to feel overwhelmed and anxious four days before the first day of your period. Or perhaps days 11–16 of your cycle are the days you feel brimming with energy, highly sociable and full of new creative ideas.

Once you've spotted your personal trends, you can begin to forward plan. Perhaps you might like to add a note to your diary to organize a night out with friends the next time you're in your sociable 11–16 stretch. Conversely, you might want to avoid any social events four days out from your period, instead setting a boundary around this time to ensure you get to slow down and rest.

Of course, you won't always be able to do this: some work, social and fitness events won't be able to fit around your cycle, and that's fine. Your cycle doesn't need to dominate your activities, but can be used as a helpful tool to shift things around where possible to help you honour your mind and body. This is a powerful and radical act of self-care, one that holds so many benefits for your physical and mental health.

Bring harmony to relationships

Take a minute to decide whether this sounds familiar: one week, you adore your partner. They look hot, all their jokes are funny and you'd quite like to hop into bed with them. Two weeks later, however, and you're wondering if they've always eaten with their mouth open, why they're sitting so close to you on the sofa, and for the love of god, why they can't just BREATHE MORE QUIETLY?! The reason you might swing between loving and hating them is – yes, you've guessed it – because of your hormones.

In the run-up to ovulation, biology wants you to have sex and conceive, hence the fact you want to be close to your partner. If you don't get pregnant though (even if you don't *want* to get pregnant), your hormones send out signals that you don't need your partner anymore, hence the fact you begin to find them, well... annoying.

Understanding the rhythms of your cycle and being able to predict when these phases are going to take place – alongside having an open and honest talk with your partner – can strengthen your relationship no end (assuming your partner is supportive and understanding). If you're able to explain that in days 23–28 of your cycle (for example), you're going to be a bit cranky/angry/emotional, and that to navigate this together you need a bit of peace and quiet, understanding and maybe a takeaway on the sofa in your PJs – and if they really hear you – it's going to bring more harmony to your relationship. Of course, revealing you'll be feeling super horny around days 13–16 might be a bonus, too! Communicating with your partner about your cycle and how it affects you can only ever work in your favour, strengthening your bond and allowing for a greater depth of understanding.

Troubleshooting: heavy/long periods

You might have never really thought about how your period compares to others', but tracking and journalling can help you realize if you fall into the category of heavy/long periods. In all honesty, you've likely realized this before, especially if you regularly need to use two forms of period protection at once or you experience "flooding" or bleeding through protection.

Heavy periods are no fun. They can lead to feelings of embarrassment and shame (even though they shouldn't) and can dominate your life for a week or so every month. Worrying about bleeding through your clothes, feeling nervous about leaving the house because your period is easier to manage in the comfort of your own home, and suffering from fatigue due to blood and iron loss can all take a toll on your life and lead to feelings of frustration and even depression.

If you have heavy periods, it's worth seeing your doctor as there's sometimes a medical cause (such as fibroids, endometriosis, hypothyroidism or a blood clotting disorder). Your doctor will be able to do blood tests and/or scans to determine the cause, and can also offer options to manage your periods, such as medication, hormonal treatment (e.g. the pill or an intrauterine system – IUS), or in some cases endometrial ablation, where the lining of your uterus is removed. Sometimes just having these causes ruled out is reassuring enough to lighten your mental load and make your period that bit easier.

In the meantime, while waiting for treatment from your doctor, you could try taking ibuprofen during your period which slows prostaglandin production, meaning less uterine shedding and therefore less bleeding.

If you struggle with severe cramping and period pain, again, don't suffer in silence. Make an appointment with your doctor and, in the meantime, try out the tips on page 36 to help manage the pain.

Troubleshooting: light/absent periods

If you always used to have a period but they've become light and irregular, or even absent, it's important to see your doctor to determine the cause. If there's a chance you might be pregnant, it's important to do a pregnancy test first to rule this out.

If you aren't pregnant, there are a few causes, including polycystic ovary syndrome (for more information, see page 54), as well as premature menopause, thyroid problems or lifestyle factors that result in amenorrhea (see pages 59–64).

The solution will differ depending on the cause, which is why a visit to your doctor is important. However, if your amenorrhea is due to lifestyle factors, there are changes you can make to help support your menstrual cycle.

As we learned on page 61, overexercising and eating too few calories are both leading causes of amenorrhea, as your body goes into starvation mode and shuts down "non-essential" processes, such as menstruation. By eating more nutrient-dense food (see page 106), and by cutting back

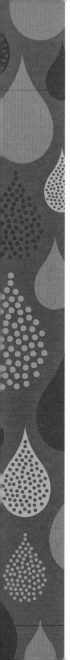

on your fitness regime, your body will begin to realize it doesn't have to conserve energy and your periods should resume. However, this can be easier said than done.

If you've been restricting your food intake or overexercising for a long time, to a point where your periods have stopped, there's a chance you may be suffering from disordered eating or an eating disorder, such as anorexia nervosa. Eating disorders are complex mental health disorders, and sometimes admitting you're struggling can seem impossible. However, amenorrhea is a big warning sign your body isn't coping. Seeing your doctor is important, as is opening up to someone you trust. There are also organizations that can offer confidential advice and support, such as Beat in the UK or the National Eating Disorders Association (NEDA) in the US. Check out our Resources on page 156 for further details.

When to see a doctor

Over the past few chapters, we've looked at various issues that might affect or impact your periods, such as medical conditions, infections and lifestyle factors including sudden weight change and stress. However, if you're still unsure whether you should see your doctor about issues relating to your menstrual cycle and periods, here's a checklist of when it's a good idea to seek medical support and advice:

- Your periods suddenly become irregular and you're under the age of 45.
- You have an irregular cycle (i.e. you have periods more often than every 21 days, or less often than every 35 days), or there's a big difference (at least 20 days) between your shortest and longest menstrual cycle.
- You're having heavy periods (for example, soaking through tampons or pads in an hour, or needing to use multiple types of period products together), and/or your periods are lasting longer than seven days.

- Your PMS symptoms are affecting your day-to-day life.
- You have irregular periods and you're struggling to get pregnant.
- You have any concerns around your menstrual cycle, periods or fertility – your doctor is there to help and support you.

Your menstrual cycle self-care plan

Now that you're more familiar with your menstrual cycle and how it impacts your energy levels, moods, appetite and more, it's time to look at how you can care for yourself to feel your best throughout your cycle.

"Self-care" is a term that's thrown around a lot these days – and one that's been adopted by the wellness industry in recent years as a way of marketing various products. However, at its most basic level, self-care simply refers to any deliberate act that protects and nurtures your own physical, mental and emotional health. It's not all scented candles and pampering – although it can include these things if they work for you. Instead, it's about more fundamental, basic acts that provide a framework for your health and well-being.

Over the following pages, we'll cover self-care practices that contribute to your overall and menstrual health, and will include:

- How to eat for optimal energy and nutrition
- Avoiding things that can make PMS symptoms worse
- Finding the right fitness regime for you, and adapting your workouts through your cycle according to your hormones and energy levels
- How to create optimum healthy sleep habits
- How to channel your anger into a productive force
- The importance of setting boundaries and saying "no"

Are you ready to make a positive difference to your life through the simple act of self-care? Then let's go…

A balanced diet for good menstrual health

A diet that's conducive to good menstrual health is also going to be good for your overall health. You probably have a good idea of what a healthy, nutritionally optimal diet looks like. But there are a few extra elements that can assist in good menstrual health…

- Healthy protein sources, such as pulses and beans, fish, lean meat and tofu.
- Wholegrain and starchy carbohydrates, such as brown rice and sweet potatoes.
- Healthy fats, such as olive oil, avocado and oily fish, including salmon (please note that oily fish should be eaten no more than twice a week if you're planning on getting pregnant someday, due to the pollutants that can build up inside your body and affect the baby's development in the womb).
- A variety of fruits and vegetables – more on this on page 108.

- Fermented foods, such as sauerkraut, kefir, kimchi and yogurt. This is because fermented foods help to regulate hormones, such as oestrogen, and can improve gut health.
- Omega-3 fatty acids, found in oily fish, walnuts, chia and flaxseed have anti-inflammatory properties that can be beneficial during your period (see page 109 for more on this).
- Treats you enjoy! Eating a balanced diet can be a wholesome and delicious experience, and the more you begin to enjoy a range of nutritionally dense foods, the easier it will be to stick to healthier eating habits. But you don't have to be nutrient-focused the whole time! A little of what you love does you good, and studies show if you allow yourself treats, you'll be more likely to stick to long-term healthy eating in the long term. A good rule to live by is 80:20 – eat well 80 per cent of the time, and allow yourself treats 20 per cent of the time.

Eat the rainbow

It's one of the first things we think of when looking to improve our diets: eat more fruit and vegetables. Recent guidelines suggest we should be aiming for eight to 10 portions daily. To help support your menstrual (and overall) health, aim to focus more on vegetables. Packed full of vitamins, minerals and fibre, and low in sugar, vegetables are brilliant all-round body boosters.

Fruit is brilliant (clearly), but it can be high in sugar, so aim for lower-sugar options. These include:

- Berries (such as strawberries, raspberries, blueberries and blackberries)
- Peaches and nectarines
- Grapefruit
- Oranges
- Melon
- Avocado

Eating a little protein with fruit can help balance your blood sugar levels, so why not incorporate some yogurt with your fruity snack, or some peanut butter with your apple slices?

Anti-inflammatory foods

When your period starts, your body releases inflammatory prostaglandins, which cause your uterus to contract and start the flow of blood. It's these contractions that cause period pain.

If you have severe period pain, you might be prescribed a hormonal contraceptive which decreases the amount of prostaglandins released and lessens the pain. However, it's thought that eating more anti-inflammatory foods could also produce a similar effect.

Anti-inflammatory foods include fruit and vegetables, wholegrains, pulses and legumes, and nuts and seeds. The spice, turmeric, also has an anti-inflammatory effect and studies have shown that it can be effective in reducing menstrual cramps. You can add turmeric to foods when cooking or take it as a supplement in the form of a capsule (many cafés have turmeric lattes on offer too, if you're feeling fancy). Ginger also has anti-inflammatory properties, so a hot mug of lemon and ginger tea might be just the thing to soothe away any PMS symptoms.

Magnesium-rich foods

Several studies have highlighted the importance of magnesium to help alleviate PMS symptoms, with research conducted in 2015 concluding that people with a magnesium deficiency were more likely to experience severe PMS symptoms, including menstrual cramping.

Snacking on magnesium-rich foods can help you reach the recommended daily intake of 270 mg. If you struggle with PMS symptoms, including some of the following in your diet each day might help:

- Nuts, including Brazil nuts, almonds and cashews
- Avocado
- Dark chocolate (containing at least 70 per cent cocoa solids)
- Legumes, including chickpeas, beans and peas
- Seeds, including pumpkin and flaxseeds
- Whole grains, such as brown rice and bread
- Bananas
- Dark leafy greens, including spinach and kale

Iron-rich foods

Iron is a vital mineral that assists with muscle, brain and immune-system function. It's stored in red blood cells and, as we learned on page 63, heavy periods can result in an iron deficiency. Therefore, keeping your iron levels stocked up through good nutrition is key.

When we think of foods rich in iron we often think of red meat; and it's true that heme iron, which is found in animal products, is more easily absorbed by the body. This means that if you eat meat, beef, dark chicken meat, liver or seafood such as shrimps are great options to keep your iron stores well stocked.

If you're vegetarian or vegan, getting enough iron is a bit trickier, but not impossible. Your body can absorb between two and 10 per cent of the non-heme iron you consume (iron from non-animal sources), so you need to eat more of it. Non-heme iron sources include dark leafy veg, lentils, beans and dried apricots. Your body finds it easier to absorb iron alongside vitamin C, so eating fruit or veg alongside your iron source is key. There are also supplements you can buy in many high-street shops, but it's recommended to consult your doctor before taking these.

The importance of cutting back on sugar

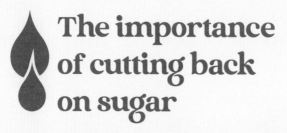

We've probably all experienced sugar cravings – and with understandable reason. Sugar is addictive, so the more you consume, the more you crave it. The trouble is, consuming excess sugar is linked to a host of health issues, including blood sugar imbalances, increased risk of diabetes and obesity, and inflammation, which we've already seen can make PMS symptoms worse. On the menstrual cycle front, excessive insulin production (a hormone your body releases to cope with high blood sugar levels) has been associated with polycystic ovary syndrome (see page 54) and impaired ovulation, because high blood sugar levels can interfere with the process of egg maturation.

Unfortunately, it's not just refined sugar (found in sweets, cakes and white bread) that causes problems. All sugar, including the natural sugars found in fruits and honey, can cause issues, too. As we mentioned on page 107, a little of what you fancy is fine, but aiming to reduce your overall sugar intake is a great way to improve your overall – and menstrual – health.

Reducing your sugar intake can seem especially difficult in the days leading up to your period, because right before it starts there's a spike in the hormone progesterone and a drop in oestrogen, which can cause your blood sugar levels to plummet (cruel, we know). This can cause you to reach for sweet treats which can be quickly broken down by the body for energy.

You can aim to break this vicious sugar cycle by ensuring you eat a range of healthy, nutrient-dense foods right through your cycle to help stabilize blood sugar levels. Try to have a range of healthy snacks to hand, such as plain popcorn, a handful of nuts, or avocado on toast to help battle cravings. And if you can't resist, opt for some squares of dark chocolate, which contains all-important magnesium (see page 110).

The importance of cutting back on alcohol

We all know we shouldn't drink too much alcohol and that we should stick to the recommended guidelines (14 units a week or below). This is because alcohol is an addictive, dehydrating depressant that is bad for your skin, hair and bank balance, and can leave you with a hangover worse than death. But what about alcohol in relation to your menstrual health?

It turns out, alcohol and oestrogen are not happy bedfellows. Many studies have revealed that alcohol increases oestrogen production. This is because when you drink alcohol, your liver prioritizes metabolizing it out of your system above all other functions, including hormone regulation. Fluctuating hormone levels due to alcohol consumption can have a range of both short- and long-term effects on your health, including:

- Increased risk of an irregular menstrual cycle
- More severe PMS symptoms
- Increased risk of breast cancer
- Increased risk of fertility issues

Some alcohols can also be high in sugar (such as rosé wine and cider) which, as we've seen on page 114, can lead to oestrogen disruption and therefore impact your menstrual cycle.

Kicking the alcohol habit or lessening your consumption can be tricky, especially if you usually drink while socializing. Alternating alcoholic drinks with sparkling water, swapping cocktails for mocktails and offering to be the designated driver on a night out can all help you stick to your low- or no-alcohol plan.

Your nutrition reflections

Hopefully, the previous pages – covering the topic of nutrition and how your diet can affect your period and menstrual cycle – have given you a little food for thought. Now, use the following journalling prompts to reflect on your relationship with food and discover where you might be able to make changes to your diet to support and improve your menstrual health.

Overall, how healthy would you say your current diet is?

_____

_____

_____

_____

When do you tend to crave sugary/unhealthy foods the most? Can you see a pattern?

_____

_____

_____

How emotional is your eating (i.e. do you only eat when you're hungry or do you sometimes eat for comfort)?

___

___

___

What is your relationship with alcohol like?

___

___

___

When making changes to your diet, you don't have to overhaul everything at once. Instead, jot down three to five small changes you could implement today to support your health (e.g. swapping sugary breakfast cereal for a poached egg on wholegrain toast).

___

___

___

___

___

Stop smoking

We're all probably aware of the very real and serious health risks of smoking. If you're a smoker, you're at a higher risk of developing lung cancer (and almost all other forms of cancer), high blood pressure, coronary heart disease, and having a stroke. And if that isn't enough to put you off, consider this: if you're a smoker, it can have a negative impact on your menstrual health, too...

- The chemicals in tobacco smoke, including nicotine, can cause hormone imbalance, which can lead to a shorter luteal phase of your menstrual cycle.
- This shortening of your cycle can affect your fertility.
- Smokers are at greater risk of having more severe PMS symptoms than non-smokers.
- Smokers are at greater risk of experiencing severe menstrual cramps than non-smokers.
- Smokers are at greater risk of early menopause than non-smokers.

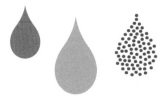

Put simply, if you're a smoker, the absolute best thing you can do for your overall and menstrual health is to ditch the habit as soon as possible!

Struggling to quit? Here are a few ideas to keep you motivated: write a list of all the reasons why quitting will improve your life (there'll be loads!). Save all the money you would have spent on smoking for a treat you really want. Use aids to help, such as nicotine patches. Ask friends and family to keep you accountable and make a plan for moments when you feel like you might cave. If you're really struggling, visit your doctor for support. There are myriad ways doctors are equipped to help you kick the habit for good, including support groups and tried and tested stop smoking tablets. Ultimately, remember that you're giving your mind and body the *best* gift by stopping smoking.

Why exercise works

It goes without saying that exercise is good for you, so it will probably come as no surprise that staying fit, active and healthy is good for your menstrual cycle, too.

We've looked at the ways in which overexercising can be detrimental to your menstrual health, potentially leading to light periods or amenorrhea (see page 61), but generally speaking, finding and sticking to a fitness regime that doesn't become obsessive or lead to extreme and/or sudden weight loss can only be a good thing for your mind and body.

In fact, many studies show that exercise can alleviate PMS symptoms, including lessened pain, cramps, bloating, mood swings and fatigue. Exercise is also known to create an upsurge in feel-good endorphins in your body, which can help to lessen feelings of irritability, sadness and depression.

You don't need to go all-out in the intensity stakes, either – when you have your period, a gentle walk, swim or some yoga are all proven to help, and are some of the best options. And with a variety of free resources available online, you can even do it from the comfort of your own home!

The best fitness workouts

When it comes to deciding what type of exercise to do to help you stay fit and active, your best bet is to choose something you enjoy. After all, if you pick a fitness regime you love, you'll be more likely to stick with it! Swimming, yoga, running, walking and cycling are all great solo activities, but if you love the camaraderie of team sports, try football, netball or hockey. Want something a little outside the box? Martial arts, circus skills and trapeze are all great for coordination as well as physical fitness. It's also a good idea to do some resistance training at least once a week, such as a circuit of bodyweight exercises (think planks, press-ups, squats, burpees and jumping jacks) or weightlifting, to help maintain your muscle mass and bone density as you age.

While all exercise is good for you, yoga could be especially beneficial when it comes to good menstrual health: one study found that people who did regular yoga classes reported fewer PMS symptoms and less period pain, as well as notable reductions in breast tenderness and abdominal bloating.

Aligning exercise with your cycle

If you look back over your tracker, you might notice your energy levels fluctuate depending on where you are in your cycle. Taking note of when you feel most and least energized, and then planning your fitness regime accordingly, can be a useful practice. It's known as "cycle syncing" and is used by some top athletes, such as female football players, to make the most of their training. If you decide to try cycle syncing, it might look something like this...

Menstrual phase

During your period, you might experience fatigue. Some gentle walks or low-impact yoga might suit you best during this time.

Follicular phase

Oestrogen levels start to rise, and so might your energy levels. This can be a good time to get stuck into endurance or resistance training, such as running, swimming and weightlifting.

Ovulation

This can be when your energy peaks, so make the most of it by scheduling some high intensity interval training (HIIT).

Luteal phase

Towards the end of this phase, you might see the return of PMS symptoms, and your energy can start to flag. You might also begin to feel irritable and/or sad. Remember, exercise is great for your mental health, so working out during this time can be key – just don't overdo it.

Remember, these are just guidelines: you don't have to cycle sync – and it might be tricky to do if you're training for an event, such as a half marathon. However, even if you need to push through certain weeks, it can be good to bear your cycle in mind and not berate yourself if you feel sluggish some weeks. Also, everyone's different: some people will feel the energy shift dramatically, while for others it might be subtle. Your energy levels might not follow the above pattern, either. Simply do what works best for you.

Your fitness reflections

Use the following journal prompts to explore your relationship with exercise and how you might begin to alter your workout regime so it aligns more closely with your menstrual cycle...

What's your current relationship with exercise like?

___

___

___

___

___

How do you (or might you) feel physically after exercise?

___

___

___

___

___

How do you (or might you) feel mentally after exercise?

___

___

___

Which workouts might suit you best at different times of your cycle? What type of exercise might you actually *enjoy* doing?

___

___

___

___

How do you feel about getting started on a new/improved fitness regime?

___

___

___

The importance of sleep

You don't need to be a scientist to know that sleep deprivation is bad news. Studies show that regular poor sleep can result in impaired memory, an increase in appetite and inflammation, lowered immunity, an increased risk of cardiovascular disease and certain cancers, and can make you more prone to diabetes and obesity... not to mention tiredness, irritability and flattened emotional responses. Fun!

Sadly, your menstrual cycle has a lot to answer for when it comes to disrupted sleep patterns. Up to seven in 10 people who menstruate say their sleep gets disrupted just before their period starts. This can mean finding it hard to fall asleep, struggling to stay asleep, nightmares, restless leg syndrome, needing to get up for a wee and – due to the drop in oestrogen, which can disrupt your body's temperature regulation

As you reach perimenopause, these sleep disruptions are only likely to increase, with some 53 per cent of people who menstruate experiencing sleep disruption at this time – and there are added struggles, too, such as night sweats.

To increase your chance of a good night's sleep, try the following:

- Get out and about in daylight hours as much as possible
- Avoid caffeine from lunchtime onwards
- Exercise in the morning, rather than the evening
- Avoid blue light (from phones and other devices) for two hours before bed
- Aim for a room temperature of 16–18°C at night
- Ensure it's dark (a blackout blind can help)
- Keep noise to a minimum – ear plugs can be a life saver, as can a white noise machine!

Rage against the machine

Before we dive into what is arguably one of the most common and well-known symptoms of PMS – anger – let's get something straight: anger is a healthy and valid emotion in the face of frustration, deceit or inequality. So if you're passionately arguing the case for something that's important to you (like how unfair it is that, due to gendered social norms, women perform the majority of unpaid care and domestic labour worldwide) and someone attempts to undermine and embarrass you by commenting, "Time of the month?", then, yes, your rage is justified. In fact, your hormones might well be simply alerting you to imbalances that exist all the time within your daily life but that during your luteal phase become intolerable.

So, if in the days leading up to your period, when your oestrogen and progesterone levels drop, you're filled with anger that you're spoken over in meetings, or no one else in your household offers to make dinner *again*, remind yourself that your anger is not only appropriate, it's justified.

Of course, feeling moody and irritable is never fun. On the next page, we'll look at some ways to channel your anger positively and constructively, and explore how to help balance out your moods. And in the meantime, stop berating yourself for feeling angry sometimes. Of course, lashing out at others with mean-spirited remarks is not kind or acceptable and requires an apology (because it's important to be accountable when we mess up), but often, there's an underlying cause for your anger that you can begin to address in a proactive, measured way.

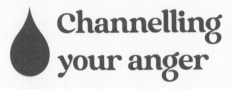

Channelling your anger

If you struggle with anger and rage when you have PMS, you're not alone. Feeling irritable and angry can be uncomfortable but, as we've seen on the previous pages, it's merely highlighting and exacerbating issues that are often there all along, whatever the time of the month.

There are things you can do to help balance out your moods. Often, simply understanding that the anger and irritability you're experiencing so strongly is partly down to hormonal changes due to your cycle can help. This is why tracking your cycle is important: it gives you the power of self-awareness, which can be calming in and of itself. Exercising regularly throughout your cycle, eating a healthy balanced diet, ensuring you're getting enough good-quality sleep and avoiding alcohol are all key when it comes to keeping your moods in check – all of which we've covered in this chapter.

However, if you don't want to (or find you can't) calm your anger but like the idea of putting it to good use instead, read on!

When used constructively, anger can steer us to respond assertively in situations where we feel we're being mistreated or taken for granted. So, are there any issues that are underlying your rage? Perhaps it's inequality in the home or workplace, perhaps you feel overlooked or unheard, or maybe you're working a bit too closely with that one colleague that rubs you the wrong way. Sometimes, these challenges feel insurmountable, but there's always something you can do, however small. It's time to be brave and channel your anger: speak up and have that difficult conversation (calmly and reasonably); sign that petition; write that letter; work from home for a day to get some space. Sometimes, self-care looks like taking responsibility, stepping up and demanding your voice is heard.

Other times, when your hormonal anger rears its head, physical activity is the only way it goes away: a sweaty workout, deep-cleaning with loud music on or going outside for a walk are all great options here as a positive way to channel that energy. Or, you might find that intentionally being quiet helps – like journalling about what's on your mind, making a to-do list or phoning a friend to vent.

Rage reflections

As we've seen, anger can be a powerful and motivating feeling. Use the following journalling prompts to help you get clear on the reasons you might be feeling rage...

Have you noticed a pattern to your anger, in relation to your cycle?

_____

_____

_____

_____

Yes, a drop in hormones is partly to blame, but what else might be underlying your anger/irritability? What do you tend to feel angry about?

_____

_____

_____

_____

What might you do to help channel this anger into something positive? Has anything specific helped in the past?

_____

_____

_____

What conversation might it be useful to have and with whom?

_____

_____

_____

What part might forgiveness play in your journey to managing your anger? What might you be able to forgive yourself for? Or others?

_____

_____

_____

The importance of protecting your time

As well as feelings of anger and irritability, PMS symptoms might also include feelings of sadness, anxiety and overwhelm (in fact, they invariably do). In order to lower your stress levels as much as possible during this period (and throughout your entire menstrual cycle in general), it's vital that you don't overschedule yourself or overpromise your time to others. In fact, protecting your own time during your luteal and menstrual phases is of utmost importance, as these are usually the phases when you need to slow down, rest and turn inwards, in order to conserve your energy as your body undertakes the task of menstruation. Using data from your tracker, can you predict when you might need to block out time for yourself over the coming months? Scheduling a little alone time might be just the thing to help with any difficult feelings you might be experiencing, such as sadness and fatigue.

How to set boundaries

If people in your life aren't used to you protecting your time, it might come as a shock when you begin putting yourself first by setting a few boundaries. But the fact is, your well-being is paramount. After all, you're no good to anyone – not your family, partner, friends or colleagues – if you're burned out by trying to do too much when your body is screaming at you to rest. Setting boundaries is imperative if you're going to carve out space for yourself during your luteal and menstrual phases.

When establishing boundaries make sure you are clear, firm and open. For example, if Thursday night drinks with the girls is a habit, you might say, "I love our nights out, but going out every week is leaving me feeling strung out, especially leading up to my period. So I'll only be joining you once or twice a month now." A calm, authoritative tone will show you're serious.

The joy of saying "no"

If you're a natural-born people pleaser and the thought of saying no to others brings you out in a cold sweat, chances are the thought of setting boundaries to protect your mental, and menstrual, health might well have you quaking in your boots.

But here's the thing: if you automatically say yes to requests without even considering your own commitments, workload, mental well-being or where you might be in your menstrual cycle first, you'll be on a fast track to feeling overwhelmed and resentful. Which is why getting comfortable with the word "no" is so important.

Sadly, in an effort to seem helpful, many of us shoulder additional responsibilities without questioning whether we want to take them on. We also often feel guilty for saying no because it can feel like a rejection of the other person. But saying no doesn't have to be rude. A simple, "Thanks for asking, but I'm afraid that won't be a good time for me," is a firm but polite response that will have you honouring your body and menstrual cycle, and will leave

you feeling empowered and with time to rest. Try not to fall into the trap of over-explaining yourself or apologizing profusely. Often, short but sweet is the best response – you're doing nothing wrong by saying "no". There really is a profound joy in saying "no" – and it's one of the best ways to ensure you work with your menstrual cycle, rather than against it.

Start putting yourself first

In this chapter, we've covered everything from tracking your cycle and following journalling prompts, to unpacking your thoughts and feelings around your menstrual cycle and coping with heavy or light/absent periods. We've also looked at self-care ideas that can help you have a more empowered and aligned cycle – such as eating well, exercising in a way that suits you, and learning how to channel your anger and protect your precious time.

Hopefully, overall, this information has helped you see that it's OK – in fact, it's necessary – to sometimes put yourself first, figure out what works for you and to then implement small, positive changes that can help you work with your menstrual cycle (and all the feelings and emotions it brings – even the "negative" ones), rather than against it.

Putting yourself first can sometimes seem like an alien concept, especially if you have caring responsibilities. However, by implementing positive changes in your life, you'll soon see that everyone benefits. By prioritizing your own health, you'll have more energy and enthusiasm to offer others in the long run. So, keep tracking your cycle if you've found it helpful, schedule in time to exercise each week, block out evenings in your diary when you know you'll need some alone time and have that conversation about dividing household chores more equally, so that you have more time to journal, rest or pursue a passion. It will all equal a happier and more empowered *you*.

Chapter Three: Breaking the Taboo

Sadly, in many parts of the world – including Western countries – there is still a stigma attached to periods. Note: if you've ever felt shame for being on your period – perhaps hiding a tampon up your sleeve on your way to a work or college bathroom, terrified someone might see it and realize you're on your period – then you, too, are a victim of the period taboo.

The good news is that in recent years things are changing for the better. Charities, activists and spokespeople are standing up for the rights of women and all people who menstruate and are instigating change. In this chapter, we'll explore why it's so vital we break the taboo around periods and how you can be part of this positive shift.

Why is menstruation stigmatized?

The authors of a powerful 2022 journal article, "The Persistent Power of Stigma"* stated that, "Sociocultural norms reinforce menstrual stigma through imperatives of concealment and hygiene, which cultivate the idea that menstruation is shameful and should therefore be hidden, controlled, and managed."

The article notes that menstruation has historically been portrayed as something unclean and abnormal. This can be dated back centuries – indeed, a Roman naturalist called Pliny the Elder believed period blood was cursed and could incite terrible events, such as causing crops to fail and wine to sour, which demonstrates just how ingrained the period taboo is. And still to this day, menstrual products are often advertised as locking in odours so that menstruators stay "fresh", perpetuating the notion that periods are dirty. This persistent negative language around periods then creates a self-fulfilling prophecy: people on their periods feel compelled to hide it, which cloaks the (completely normal) process as something to be ashamed of, which makes people feel dirty... and so the cycle continues.

*Olson, Mary M *et al*. "The persistent power of stigma: A critical review of policy initiatives to break the menstrual silence and advance menstrual literacy." *PLOS global public health* vol. 2,7 e0000070. 14 Jul. 2022.

Periods and the education gap

With all this cultural conditioning that periods are something to be ashamed of, you might not be surprised to hear that having your period affects school attendance worldwide – something that is fundamentally *not* OK.

According to the charity Action Aid, girls often miss one or more days of school when they have their period. Girls in Sub-Saharan Africa were found to miss as much as 20 per cent of their school year – something that can have devastating consequences, as lack of education can mean girls get forced into child marriage.

In the UK, too, there is an alarming rate of school absence due to periods. In a survey conducted by the charity Plan International UK in 2021, it was found that nearly 2 million girls (64 per cent) aged 14–21 have missed a part day or full day of school because of their period, with 13 per cent missing an entire day of school every month. It blamed a "toxic trio" of issues: lack of proper education around periods; the stigma and shame that surrounds menstruation; and the cost of period products.

With one in six girls saying they've been teased or bullied for having their period, it's clear there's still a long way to go to break the taboo – and that breaking the taboo is vital if the next generation of girls and all people who menstruate are going to grow up feeling empowered and proud of their bodies.

What is period poverty?

Everyone who menstruates deserves to feel comfortable and have access to appropriate period protection, such as pads, tampons or period underwear. Unfortunately, though, this is not always the case.

The UK charity, Bloody Good Period, found that in 2022, 24 per cent of people who menstruate in the UK struggled to afford period products. The menstrual movement PERIOD reported that, in 2023, nearly one in four students struggled to afford period products in the United States. A US study published in 2021 found that period poverty impacted mental well-being, with those affected more likely to experience moderate to severe depression. Can you see the link? Imagine knowing that every month your period would come, and every month you wouldn't be able to buy the basic items you need to get through it. It's no wonder this stress-inducing situation negatively impacts mental health.

Period poverty is a huge problem among those who are marginalized in society, including refugees, asylum seekers and homeless people. Thankfully, there are charities working hard to address this. In 2022, Bloody Good Period provided more than 119,000 packs of period products to those who could not afford them in the UK, including refugees; in the US, the organization PERIOD is raising $250,000 in order to send millions of disposable and reusable period products to people in need. To find out how to help, including how to donate and fight against period poverty, see page 151.

All is not lost

With all this talk of shame and stigma, you might imagine the future is bleak for people who menstruate. But all is not lost – we *are* taking steps in the right direction.

In recent years, charities, campaigners and activists have started bringing conversations about periods into the mainstream to help break the taboo and start dismantling generations of taboo and stigma. The language around periods is changing as more people begin to realize that words matter (see page 149 for more on this). Books, articles and podcasts are now normalizing periods – and normalizing talking about periods – such as Maisie Hill's fantastic books *Period Power* and *Perimenopause Power,* and the Radio 4 podcast *28ish Days Later*.

What's more, all this brilliant activism is helping bring about much-needed change on a systemic level. In 2021, the UK government abolished its "tampon tax" – the VAT charge added to the cost of single-use period products, making them more affordable. And in January 2024, after a two-year campaign, this was finally extended to period underwear, too, to make them more cost-effective and encourage people to make the switch to a more environmentally friendly and sustainable choice. The "tampon tax" has been abolished in a number of other countries too, including Canada, India, Kenya, Mexico and South Africa.

It's time to tackle period stigma

Breaking the period taboo is important. Not only for ourselves and for future generations, but also for the teen who lives down the road who can't afford period products, for the girl who has been forced to leave education in Sub-Saharan Africa, and for both the cis and transgender communities. Only by having frank and open conversations about periods – with our families and friends, including the boys and men in our lives – can we begin to pull apart decades of learned shame and stigma.

While progress has been made, there is still work to do – and much of it can be done simply by having conversations using frank language and by supporting those around us. Here's how you can take some small steps to help instigate positive change...

Don't be afraid to discuss periods

When talking about periods, words are important. Always use frank language such as "periods" and "menstruation" rather than euphemisms such as "time of the month", "having the painters in" or "Aunt Flo is visiting". Euphemisms only perpetuate the myth that periods are something shameful that should be cloaked in secrecy, rather than something it's OK to talk about openly.

Language around period products is also important. "Sanitary products" and "sanitary wear" should be swapped for "period products/supplies" or "menstrual products/supplies", for the simple fact that "sanitary" suggests periods are something dirty that need to be cleaned up. Likewise, "feminine hygiene products" should be avoided because not only is there absolutely nothing unhygienic about having a period, but also periods are not a mark of femininity for all – period language needs to be inclusive. Which leads us to…

Use inclusive language

Just as not all women have periods (due to, for example, menopause, PCOS or surgeries such as a hysterectomy), not everyone who has a period is a woman. As we mentioned at the start of this book (page 9), however a person identifies, if they have periods then they have the right to support, advice and information. However, if this support doesn't sound like it's for them because it only talks about periods in the context of girls and women then it becomes exclusionary and can feel inaccessible. When talking about periods, make sure your language is inclusive – "people who menstruate" or "women and people who menstruate" ensures everyone feels seen, included and supported. If you struggle with this, please remember, this is not in any way about erasing women from the conversation. The positive period movement is still very much about championing women and women's rights – but it's also about championing the rights of trans and non-binary people, too. By using inclusive language, everyone is included in the conversation.

Donate period products

Why not consider donating period products to those who can't afford them? The Trussell Trust, which runs nationwide food banks in the UK, states that while most people donate food, non-food items also rank high on their list of essentials, with period products like pads and tampons in great need. Many supermarkets and other retail outlets now have food bank collection points, so it's easy to pick up an extra pack of pads or tampons while you're shopping and donate them. The Hygiene Bank accepts donations of period products alongside items such as toothpaste and shower gel to help end hygiene poverty. In the US, there are various projects, such as The Period Pantry , where you can donate products via an Amazon Wish List. Details of their websites are on page 156.

Donating period products is quick and easy, and could make someone else's life that bit easier and safer.

Request free period products in workplaces

The UK's period product scheme now provides free period products to students aged 16–19 in their place of study. Many workplaces, too, now recognize the importance of placing free period products in bathrooms for staff. Does yours? If not, sending a polite email to HR making this perfectly reasonable request is a great start. Often, change arises from people seeing a problem and then taking those first small steps to fix it – pointing out an oversight such as this could make a huge difference for people in your workplace who can't afford period products, have forgotten to bring them with them, or who have come on their period unexpectedly.

Support period charities

Supporting menstrual charities is a great way to make an impact. There are so many amazing charities, organizations and movements out there that are taking big strides forward in the battle to end period poverty and put a stop to period stigma for good.

Charities such as Bloody Good Period and Period Poverty in the UK help to get pads to those who need them, including homeless people, refugees and asylum seekers, as well as working to normalize periods; while Binti is a period charity that runs projects in the UK, India, the Gambia and America, and is working towards a world in which everyone has menstrual dignity.

By adding your voice to an established movement, you're helping promote positive change – why not join their cause? Their website details are on page 156.

Conclusion

Incorrect and outdated societal views about periods have led to many of us spending (sometimes decades) feeling ashamed of our bodies. We've unconsciously internalized the stigma and taboo surrounding our menstrual cycles, and as such have felt the need to stay silent about our periods – sometimes to the detriment of our own health and well-being.

But hopefully this book has now given you the confidence to say, "Enough is enough!"

After becoming familiar with your own cycle through the tracking and journalling exercises within these pages, as well as looking at ways you can work *with* your cycle rather than against it, you can now move forward feeling empowered and proud of your (frankly amazing) body. If you haven't already, it's time to shrug off any shame and stigma you've been carrying – your menstrual cycle is incredible and something to be celebrated!

Resources

Websites

Beat (beateatingdisorders.org.uk) – for support with eating disorders and disordered eating if you need help with amenorrhea

Binti (bintiperiod.org) – international organization working towards menstrual dignity for all

Bloody Good Period (bloodygoodperiod.com) – health education to normalize periods and end period poverty

Endometriosis UK (endometriosis-uk.org) – information, support and advice if you're living with endometriosis

Eve Appeal (eveappeal.org.uk) – UK charity funding research and raising awareness into gynaecological cancers

International Association for Premenstrual Disorders (iapmd.org) – support, information and advice for people living with premenstrual dysphoric disorder and premenstrual exacerbation

LGBT Foundation (lgbt.foundation) – support, advice and advocacy for those in the LGBTQ+ community

National Eating Disorders Association (nationaleatingdisorders.org) – leading eating disorders organization in the US

Period (period.org) – working to end period poverty in the US

Period Poverty (periodpoverty.uk) – working to eliminate period poverty in Britain's most deprived areas

Spectra London (spectra-london.org.uk) – UK-wide support for LGBTQ+ people, including peer-led trans services
Verity (verity-pcos.org.uk) – self-help support group for people living with polycystic ovary syndrome

Books
Dealing with Problem Periods by Dr Anita Mitra (2024)
Menopausing by Davina McCall and Naomi Potter (2022)
Period by Emma Barnett (2019)
Period Power by Maisie Hill (2019)

Podcasts
28ish Days Later – intimate, taboo-busting series
FLOW – demystifies and destigmatizes heavy menstrual bleeding
PERIOD – explores anything and everything to do with the menstrual cycle
Period Story – facilitates taboo-breaking conversations about periods
That's On Period – the cofounders of Period Poverty Project in the US debunk social stigma
The Maisie Hill Experience – insights and strategies to help improve your cycle

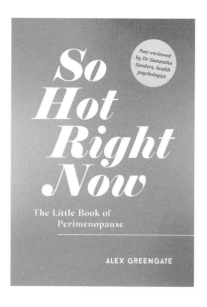

So Hot Right Now
The Little Book of Perimenopause

Alex Greengate

Paperback

ISBN: 978-1-80007-709-6

Take charge before the change

The years leading up to the menopause can be a daunting time, and one which is widely misunderstood. Fear not! This book is here to break the stigma and share the knowledge, answering the questions you've been too afraid to ask and demystifying the perimenopause once and for all.

Filled to the brim with essential information, this book will take you through all the stages and symptoms of the perimenopause right up to the menopause, so you can face this new stage of life with confidence.

Strong Woman Era

How to Embrace Your Strength and Elevate Your Life

Saffron Hooton

Hardback

ISBN: 978-1-83799-492-2

Empower your mind, strengthen your body and build your best life with this girl-power guide to strength training.

This is your *Strong Woman Era*, a new phase of growth, fulfilment and success. It's time to invest in practices that make you feel mentally and physically strong – because you are! Through resistance training, you can harness your power and realize just how much you can achieve. That's where this guide comes in.

Filled with helpful tips, practical advice and energizing affirmations, this wellness and weightlifting companion will support you on your journey to becoming your strongest self.

Have you enjoyed this book?
If so, why not write a review on your favourite website?

If you're interested in finding out more about our books, find us on Facebook at **Summersdale Publishers**, on Twitter/X at **@Summersdale** and on Instagram and TikTok at **@summersdalebooks** and get in touch. We'd love to hear from you!

Thanks very much for buying this Summersdale book.

www.summersdale.com

Image Credits

Blood drops throughout © kidstudio852/Shutterstock.com
p.10 – female reproductive system diagram
© Fandorina Liza/Shutterstock.com
pp.29, 35, 71, 75, 79 – period products © Iris vector/Shutterstock.com
p.41 – food © GoodStudio/Shutterstock.com
p.46 – phone © Mark Rademaker/Shutterstock.com
p.103 – stethoscope and clipboard © ichico/Shutterstock.com
p.145 – money jar © Nite Studio/Shutterstock.com